START WITH A QUESTION

# Are Icebergs Salty?

## FAQs About Earth

Cynthia O'Brien

A Crabtree Forest Book

Crabtree Publishing
crabtreebooks.com

# Notes for Readers and Class Discussion

This book is designed to teach readers about core subject areas, build curiosity, and inspire further investigation. Readers are encouraged to build upon what they already know about the subject and engage in topics they want to learn more about. Here are a few guiding questions to prompt reflection and discussion. Possible answers appear in red.

## Before Reading:

*Read the title and look at the table of contents. What do you already know about Earth?*

- *I know that humans must work together to protect Earth.*
- *I know that Earth has a huge variety of ecosystems with many plants and animals.*

*What would you like to learn about Earth?*

- *I would like to learn about unique plants and animals that live on Earth.*
- *I would like to learn about what lies beneath Earth's surface.*

## During Reading:

*Pause after reading each page or chapter. What questions do you have about what you read? What are you curious to know?*

- *I wonder how Earth was created.*
- *I am curious to know what plants and animals can survive in the deepest parts of the ocean.*

*Make connections to what you are reading. How is this like something you already know?*

- *I know that natural disasters can cause an extreme amount of damage.*
- *I know that people do not have equal access to Earth's resources, including food and safe drinking water.*

## After Reading:

*Recall key details about the book. What was the author trying to teach readers?*

- *The author was trying to teach readers that scientists ask questions to learn about a topic.*
- *The author was trying to teach readers that humans need to take steps to slow climate change.*

*How did the images and captions help you understand more?*

- *The photographs helped me see what scientists have discovered about Earth's features on land, in air, and in water.*
- *The captions helped me understand new details about Earth.*

# Table of Contents

# Think Like a Scientist

When you want to learn about something, what is the first thing you do? If you begin by asking questions, you are thinking like a scientist. Scientists ask questions to investigate all subjects, from what lies beneath Earth's crust to why living things behave in certain ways. Scientists carry out experiments to investigate their questions. They draw conclusions from the results of the experiments. Sometimes, the answers lead to even more questions.

Earth is one planet in the Milky Way **galaxy**. It is the only planet that we know has liquid water on its surface and supports life. Earth has fiery volcanoes, massive oceans, and towering mountains. To learn about Earth, there are all kinds of questions you can ask. For example, where is the deepest lake? Why do earthquakes happen? How fast does a tornado travel?

In the pages that follow, you will discover some amazing facts about Earth. The information you learn may even spark your curiosity and inspire you to ask more questions.

*A physical map of Earth shows different landforms, deserts, and seas.*

Earth is not a perfect sphere. It bulges around the **equator** due to how it rotates.

# How old is Earth?

Earth is around 4.5 billion years old, but scientists do not know its exact age. In fact, they estimate that this number could be off by about 50 million years. In the 1660s, Danish scientist Nicholas Steno was the first to realize that layers of **sedimentary** rock (see p. 8) could tell us about Earth's history. Steno explained that the oldest layers were on the bottom and the newest layers were on the top. This was a major discovery, but still gave no clear idea about Earth's age.

In 1863, William Thomson, an Irish mathematician and physicist, calculated that Earth was between 20 million and 400 million years old. Thomson, who became Lord Kelvin, based this age on an estimated time it took Earth to cool after it formed.

In the 20th century, scientists began to date rocks using a radiometric method. This process measures the rate of decay, or breakdown, of **radioactive elements** within a rock. But there are still some questions about this method. For example, did the speed of the decay stay the same?

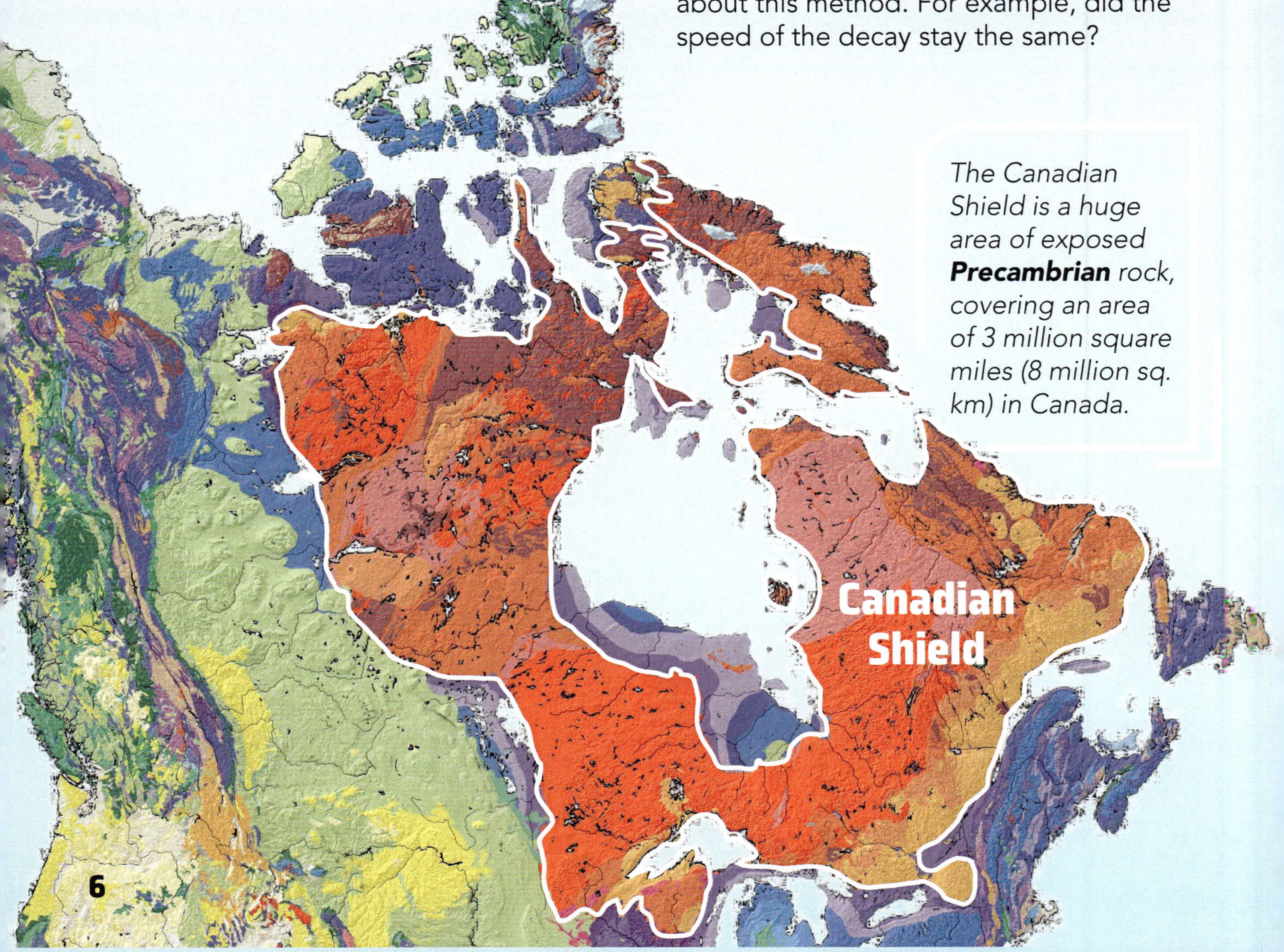


*The Canadian Shield is a huge area of exposed **Precambrian** rock, covering an area of 3 million square miles (8 million sq. km) in Canada.*

So far, scientists have found Earth's oldest rocks in northern Québec, Canada. These rocks, part of the Canadian Shield, are 4.28 billion years old. Tiny mineral grains found in the Jack Hills of Western Australia, however, are older still. They are around 4.3 to 4.4 billion years old.

*Strata are defined beds, or layers, of sedimentary rock—shown here, in Coyote Buttes, Arizona.*

# What is Earth made of?

Earth is made up of four layers: an inner core, an outer core, a mantle, and a crust. The inner core is a hot, solid mix of iron and nickel. A thick outer core, mostly made up of liquid iron and nickel, surrounds it. The mantle is a hot layer of solid and molten, or liquified, rock. Finally, the solid crust covers the mantle.

Earth's crust is made of igneous, metamorphic, and sedimentary rock. Igneous rock, such as granite, is made from cooled magma. Metamorphic rock, including marble, has been changed by pressure and heat. Sedimentary rock, such as limestone, forms from eroded rocks, deposits of sediment, and once-living **organisms**.

In the 1800s, scientists thought Earth's core was liquid and surrounded by a solid mantle and crust. A Danish scientist, Inge Lehmann, looked at the **shock waves** caused by an earthquake in New Zealand in 1929. By studying how the waves traveled differently through solids and liquids, Lehmann proposed in 1936 that Earth's center consisted of two parts, a solid inner core and a liquid outer core. Later studies confirmed Lehmann's theory.

*Magma is very hot, fluid rock under Earth's surface. Much of it is in the mantle. When volcanoes erupt, magma spews out as lava.*

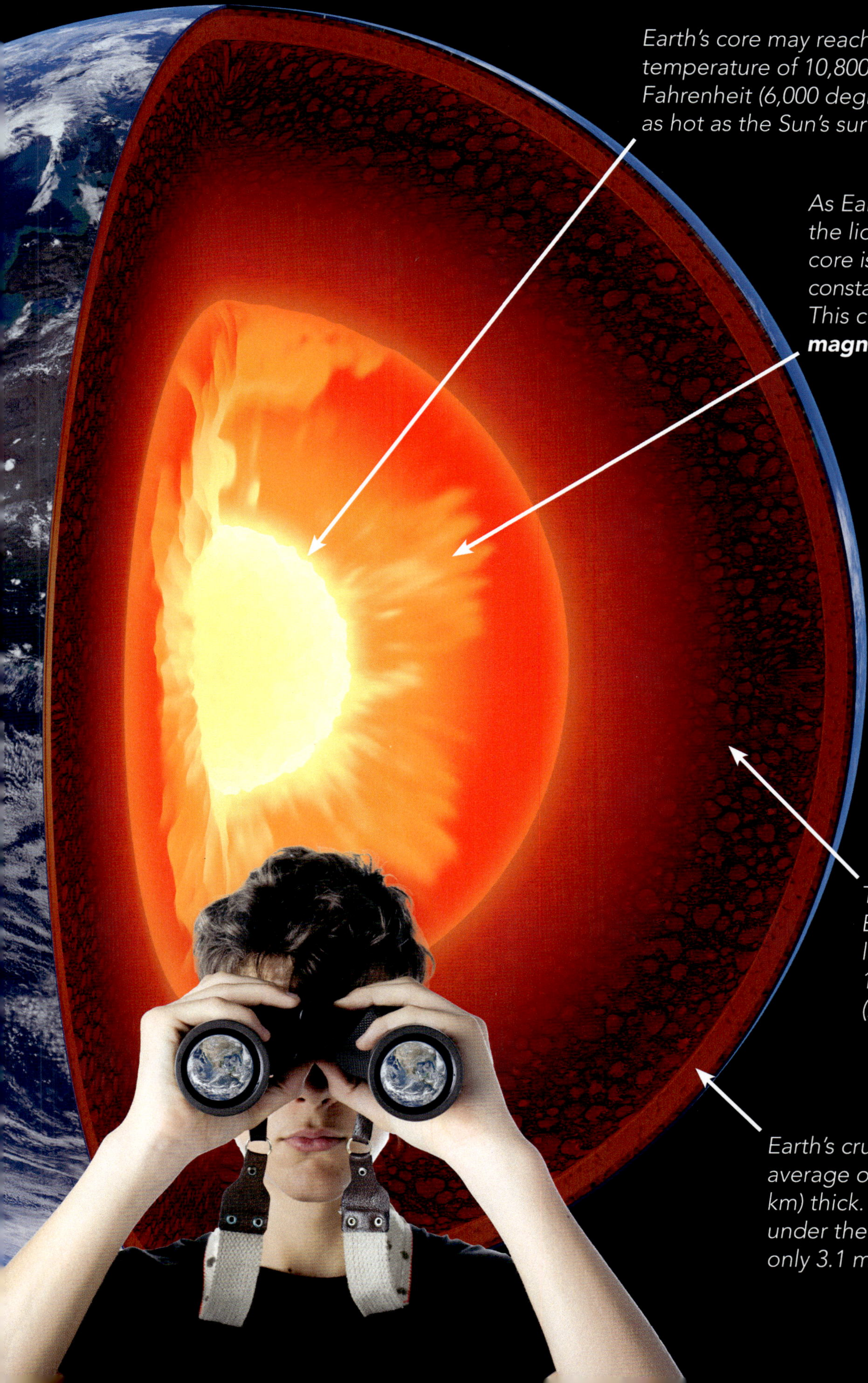

Earth's core may reach a temperature of 10,800 degrees Fahrenheit (6,000 degrees Celsius), as hot as the Sun's surface.

As Earth rotates, the liquid outer core is also in constant movement. This creates Earth's ***magnetic field***.

The mantle is Earth's thickest layer. It is about 1,800 miles (2,900 km) thick.

Earth's crust is an average of 19 miles (30 km) thick. It is thinnest under the oceans, at only 3.1 miles (5 km).

# How did the continents form?

Earth's continents started to form around 4 billion years ago. But they were not the continents we see today. Instead, there were large areas of land along or south of the equator. About 240 million years ago, most of this land joined together to form one continent, called Pangea. Over millions of years, Pangea broke apart and spread out to form today's continents.

All of this change was caused by the **tectonic plates** that make up Earth's crust. There are seven large plates, as well as several smaller ones. The plates are floating on Earth's mantle, constantly shifting at about the same speed that fingernails grow. These tectonic plates move apart, slide under one another, move past each other, or collide.

**Before**

**After**

*The movement of tectonic plates creates fault lines, or cracks, at the plate boundaries. The North American and Pacific plates touch and slip past each other along an 800-mile (1,287 km) boundary called the San Andreas **Fault** in California.*

# Why do earthquakes happen?

Most earthquakes are caused by the movement of tectonic plates at fault lines. When the plates shift and edges get stuck, they create **friction**. The plates push against each other, building up pressure and energy at an **epicenter**. Then, the plates suddenly move past each other, releasing the energy as shock waves. The waves cause the ground to shake, sometimes sending vibrations hundreds of miles away. Volcanoes and underground mine explosions can also cause earthquakes, although these are usually less powerful.

## DID YOU KNOW?

**About 90 percent of the world's earthquakes take place in an area called the Ring of Fire in the Pacific Ocean. It stretches 24,900 miles (40,000 km) and crosses the boundaries of several tectonic plates.**

*The Ring of Fire also contains about 75 percent of all Earth's volcanoes.*

# Where is the biggest cave on Earth ?

The Son Doong (Mountain River) cave in Phong Nha-Kè Bàng National Park, Vietnam, is about 2.5 million years old and over 5.5 miles (8.9 km) long. Several areas are more than 328 feet (100 m) wide and 656 feet (200 m) high. The cave formed when two large rivers joined together and carved a huge tunnel into the limestone mountains.

The cave system is even home to two underground forests! These were created when the cave ceiling collapsed above two large sinkholes, letting in rain and light.

*Son Doong cave contains massive rock formations such as the Hand of Dog* ***stalagmite*** *(above). It is more than 230 feet (70 m) tall.*

*Son Doong cave is so large that it could fit several 40-story skyscrapers!*

*Kīlauea erupts often—every two or three years on average.*

# How hot is a volcano?

The temperature inside a volcano depends on the **composition** of the magma, or molten rock. It can range from 1,300 to 2,200 degrees Fahrenheit (700 to 1,200 degrees Celsius). Kīlauea, in Hawaii, emits the most energy, with a temperature of about 2,140 degrees Fahrenheit (1,170 degrees Celsius).

Kīlauea's eruptions are so hot because they are effusive. This means that lava flows steadily out rather than exploding. The magma in an effusive eruption is hotter because it comes from deeper under Earth's surface.

## DID YOU KNOW?

**Mauna Loa in Hawaii is the world's largest active volcano. Active volcanoes have erupted in recent history and may erupt again. Mauna Loa extends 13,681 feet (4,170 m) above sea level and 16,319 feet (4,974 m) below sea level, making it 30,000 feet (9,144 m) tall. This is taller than the highest mountain, Mount Everest, at 29,035 feet (8,850 m)!**

# How much water is on Earth?

Water covers about 71 percent of Earth's surface. Around 96.5 percent of this water is in the oceans. About 1.7 percent is fresh water frozen in glaciers and ice caps, and less than 1 percent is fresh water found in surface water such as lakes and rivers. Another 1.69 percent is groundwater. Around half of that is fresh water and the other half is salt water. Lastly, there is a tiny percentage of water in Earth's **atmosphere**.

In total, there are more than 332,519,000 cubic miles (1,386,000,000 cubic km) of water on Earth. This is about 352,670,000,000,000,000,000 gallons (1,335,001,173,863,280,128 kiloliters)! If we could put all Earth's water in a bubble, it would be about 860 miles (1,385 km) across.

Oceans have salt water—but the amount of salt changes between them. The Atlantic Ocean contains the most salt of all the oceans. The salt in ocean water comes from rocks on land, undersea vents, and underwater volcanoes. Rain breaks down rock, carrying salt and minerals along rivers and to the ocean. Vents and volcanoes allow large amounts of dissolved minerals to enter the ocean water.

*The Red Sea, which is part of the Indian Ocean, contains the saltiest, hottest water of all. It is hot due to an undersea **trench** and volcanic activity.*

The Pacific Ocean is the largest and deepest of all the oceans. It contains the Mariana Trench, which is 36,000 feet (10,973 m) at its deepest.

The Sun evaporates ocean water to leave behind sea salt.

There are millions of lakes on Earth. Most contain fresh water, but not all. An open lake has rivers and streams flowing from it. These lakes contain fresh water. A closed lake does not have water that flows in or out. It can be salty because water evaporates and leaves salts behind. The Dead Sea is a salt lake that lies in a desert between Jordan and Israel.

Lakes can be formed in different ways. Some, such as the Great Lakes in North America, were formed when slow-moving glaciers carved basins into the earth. These basins filled as the glaciers melted. Other lakes, such Lake Baikal, in Russia (see page 21), formed when shifting tectonic plates created faults in Earth's surface that filled with water from rivers and rain. Volcanoes, rivers, and landslides can also form lakes. For example, after a volcano erupts, it leaves a sunken area in its center called a caldera. If the volcano remains inactive, rain fills the caldera and creates a lake.

*The very salty water in the Dead Sea is **denser** than the human body, making it very easy to float!*

*The Dead Sea is too salty for most organisms to survive, though some bacteria live there.*

*The Great Lakes, which lie between Canada and the United States, contain about 18 percent of the world's fresh surface water.*

*Though some survivors hope to return, it is still unsafe to live near Lake Nyos.*

## DID YOU KNOW?

**Lake Nyos, in Cameroon, is one of a few lakes on Earth that explode. These natural disasters are called limnic eruptions. They may happen due to events such as earthquakes and volcanic activity, and high amounts of carbon dioxide in the water. When eruptions occur, a deadly fog of carbon dioxide can be released. When Lake Nyos erupted in 1986, 1,746 people and thousands of livestock were tragically killed.**

*Large icebergs may be as high as 300 feet (90 m) above the ocean's surface. But about 90 percent of an iceberg lies below the surface.*

# Are icebergs salty

Icebergs are large chunks of ice that float in cold, salty oceans, but they are made of fresh water. That's because the ice breaks off from glaciers or ice shelves in a process called calving. Glaciers are slow-moving masses of ice. Ice shelves are floating platforms of ice that extend from a glacier into the ocean.

Cracks in glaciers and ice shelves develop over time, causing chunks of ice to break off at the edges. Icebergs can also break apart into smaller pieces known as bergy bits and growlers. Most icebergs have two main shapes. **Tabular** icebergs have flat tops and steep sides. Non-tabular icebergs may have round, pointed, or other irregular shapes.

# How fast does a glacier move?

Glaciers move slowly—about 1 inch (2.5 cm) per day. A surging glacier, however, may move over 328 feet (100 m) in a day. This kind of movement can happen suddenly, usually after long periods of ice buildup and slow movement.

Glaciers can move forward or backward. When more snow is added and less melting or evaporation occurs, glaciers will advance. If less snow is added and more melting occurs, glaciers will retreat.

## DID YOU KNOW?

**Some glaciers are more than 2 million years old! Most glaciers formed during the most recent ice age. An ice age is a long period in which ice covers large areas of land. Earth has had five major ice ages. The first started about 2 billion years ago and lasted about 300 million years. The fifth began about 2.6 million years ago and ended about 10,000 years ago.**

*The Antarctic ice sheet is a glacier that covers 5.4 million square miles (14 million sq. km). This is almost twice as big as Australia.*

# Where do rivers begin?

The source, or beginning, of a river is called its headwater. Often, rivers begin high up in mountains from smaller rivers made from rain and melting snow. Other rivers start at the edge of a lake or from water underground that rises to the surface. Rivers flow into oceans or into lakes. The end of the river is called its mouth.

It is not always easy to find a river's source. The Amazon River in South America is one of the world's longest rivers. Explorers are still searching for its true source.

*The Amazon begins somewhere in the Andes Mountains and flows at least 4,000 miles (6,400 km) to the Atlantic Ocean.*

Lake Baikal is the world's oldest freshwater lake. It formed about 25 million years ago.

# Where is the deepest lake on Earth?

Earth's deepest lake is Lake Baikal in Siberia, Russia. The lake is about 5,600 feet (1,700 m) deep, or almost as deep as the Grand Canyon at its deepest point. It contains about 20 percent of Earth's unfrozen fresh water.

More than 330 rivers and streams flow into the lake, and it is surrounded by high mountains. Lake Baikal stretches about 395 miles (640 km) and has an average width of about 30 miles (48 km). It contains around 45 islands.

*More than 2,600 plant and animal species live in and around Lake Baikal. Around 1,000 of them, such as the nerpa, or Baikal seal, are not found anywhere else in the world.*

Burning meteoroids become meteors, or shooting stars, that are visible on Earth.

## CHAPTER 3: QUESTIONS ABOUT THE ATMOSPHERE

# What is air made of ?

Air is made mostly of gases. Each layer of Earth's atmosphere—the troposphere, stratosphere, mesosphere, thermosphere, and exosphere—has different combinations of gases.

The troposphere extends from Earth's surface to about 7 miles (11 km) above Earth. It is made up of about 21 percent oxygen, 78 percent nitrogen, and 1 percent other gases and particles. The troposphere is where most weather occurs and it contains the air living things need to survive. The stratosphere is about 7 miles (11 km) to 31 miles (50 km) above Earth. The stratosphere contains the ozone layer, which protects Earth from the Sun's harmful **radiation**. The third layer of Earth's atmosphere is the mesosphere. This layer is about 22 miles (35 km) thick and protects Earth from **meteoroids** by burning them up before they reach Earth.

Thermo means hot, and the temperature in the thermosphere can reach 4,500 °F (2,482 °C). The International Space Station orbits Earth in this layer. The final layer of the atmosphere is the exosphere. It contains gases such as hydrogen and helium and extends between 440 miles (708 km) and 6,200 miles (9,978 km) above Earth.

Many experts use the Kármán line (see p. 25) to define where outer space begins.

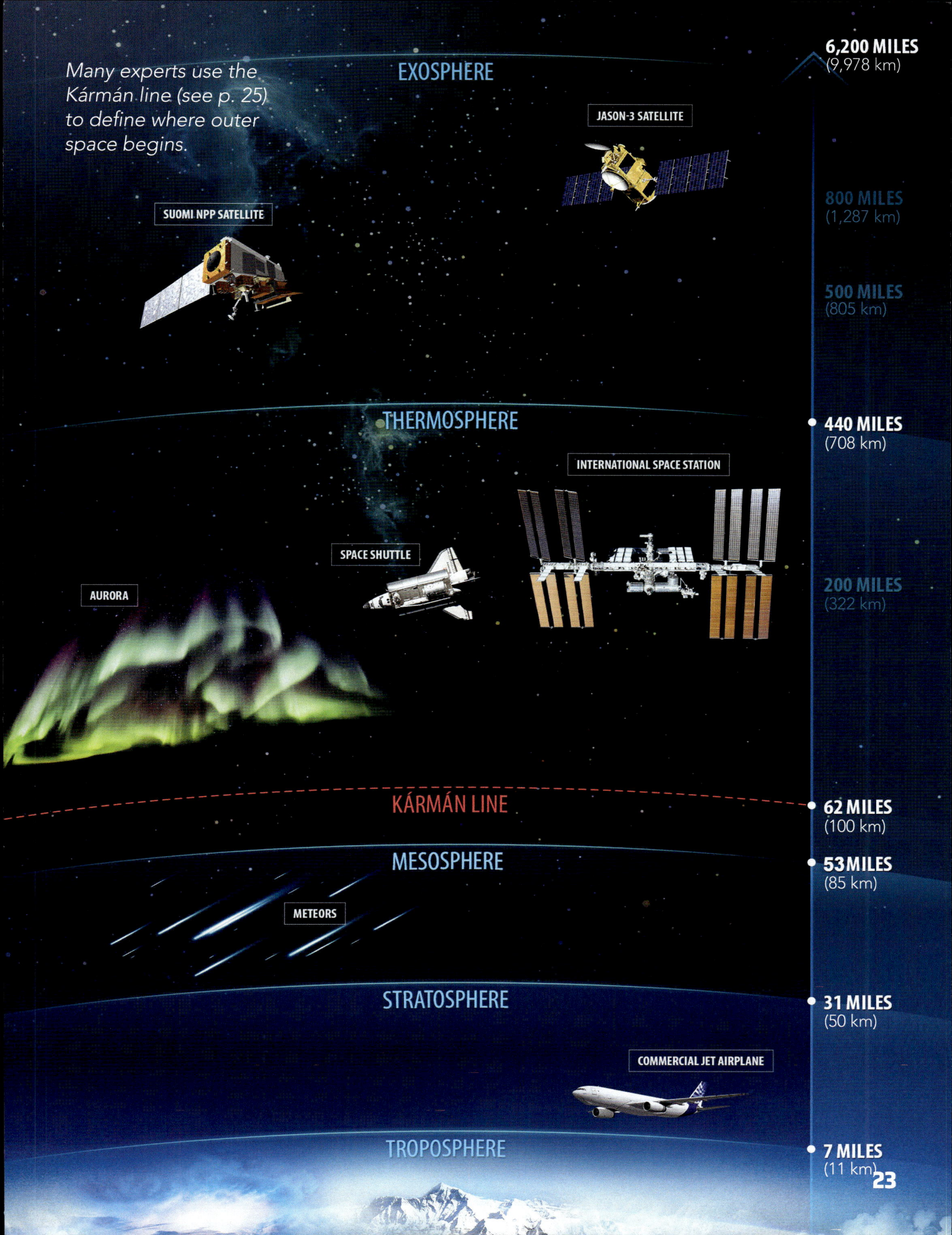

Most of Earth's satellites orbit in the exosphere.

*Generally, planes cannot fly above the Kármán line. But since 2004, some companies, such as Virgin Galactic, have launched special planes that go beyond the line.*

Earth's atmosphere does not end abruptly at the upper part of the exosphere. It becomes thinner and thinner the farther away it is from Earth, and gradually fades away. In 2019, NASA and the European Space Agency (ESA) used information from a space observatory to learn more about Earth's outer atmosphere. They discovered that Earth's **geocorona** stretches 391,000 miles (629,300 km) above Earth. This means that Earth's atmosphere could extend far beyond the Moon and its orbit.

It may sound surprising that Earth's atmosphere and what some scientists consider to be "outer space" overlap. These scientists use the Kármán line, an imaginary boundary, to define where space begins, however. Not all scientists agree where this line falls or that this is where space really begins. Usually, the Kármán line is said to be 62 miles (100 km) above sea level on Earth. This actually falls in the lower thermosphere, meaning it is within Earth's atmosphere.

## DID YOU KNOW?

**Earth's atmosphere has not always had oxygen. When the planet first formed, its atmosphere was a mixture of other gases such as carbon dioxide. The Sun helped change the atmosphere, leaving only nitrogen and carbon dioxide. Then, more than 3 billion years ago, tiny organisms evolved that used the Sun, water, and carbon dioxide to grow and produce oxygen.**

*In 1972, astronauts on the Apollo 16 Moon mission took the first photograph of the geocorona.*

# What causes the northern lights?

The aurora borealis, or northern lights, are colorful spectacles that occur near Earth's North Pole. There are also southern lights, or aurora australis, which appear near the South Pole. The auroras happen about 80 miles (129 km) above Earth's surface or higher.

Auroras are caused when **solar wind** becomes trapped in the ionosphere, a part of the thermosphere. Solar wind is made up of ions, or electrical particles from the Sun. Earth's magnetic forces protect the planet from most of the solar wind, but when it does break through, it becomes trapped for a short time in the ionosphere. There, the solar wind collides with oxygen and nitrogen to create a colorful glow.

*The colors of the auroras depend on the atoms the ions crash into and their altitude, or height above Earth's sea level. Yellow-green colors mean the ions struck oxygen at a lower altitude.*

# Are clouds heavy?

Fluffy cumulus clouds look like lightweight tufts of cotton. In reality, a typical cumulus cloud weighs about 1.1 million pounds (500,000 kg), or about the weight of 350 cars! Clouds are heavy because they are made of water droplets or ice crystals.

There are many kinds of clouds that form above Earth's surface. Stratus clouds are low and cover the sky in gray sheets. Cumulonimbus clouds are heavy, low clouds with a dark bottom. Higher up are feathery cirrus clouds.

**DID YOU KNOW?**

**When cumulonimbus clouds form in the winter, they can cause a rare event called thundersnow. This is a snowstorm with thunder and lightning. Scientists are still learning more about how these storms happen.**

Cumulonimbus clouds produce thunder, lightning, and rain. They can also produce thundersnow and hail.

# Where does the wind come from?

Wind comes from differences in air pressure, or the weight of gases in the air. The Sun warms some areas of Earth more than others and creates different air pressure systems. When warm air rises, an area of low pressure is created. When cool, dense air lowers, a high-pressure area is created. Wind occurs when air from a high-pressure system rushes in to fill the space left by the rising air in a low-pressure system.

Changing air pressure can also cause different wind patterns. Trade winds blow from east to west near the equator. Westerlies are powerful winds that blow from the west. Katabatic winds occur when gravity pulls higher, dense air down.

*Antarctica is the windiest place on Earth. Its katabatic winds can reach 200 miles per hour (320 km/h).*

*Hurricanes form over the ocean and have spinning wind speeds between 74 miles per hour (119 km/h) and over 157 miles per hour (252 km/h).*

# How fast does a tornado move?

Tornadoes are fast-moving spirals of air that form during large thunderstorms. A tornado can travel at more than 60 miles per hour (97 km/h), but most move about 10 to 20 miles per hour (16 to 32 km/h).

Scientists estimate that wind speeds within a tornado are more than 300 miles per hour (483 km/h). But it is impossible to get a true measurement of wind speeds inside a strong tornado because the weather equipment would fall apart.

*Tornadoes can be deadly weather events that cause great damage. They can lift objects as heavy as cars, and carry objects as far as 100 miles (161 km) away.*

# How does animal extinction affect an ecosystem?

Earth supports a huge variety of life. Living things interact and exist with each other within Earth's many ecosystems. They need each other to survive. Each ecosystem provides the ideal environment for its animals and plants to survive.

Biodiversity, or the variety of life on Earth, keeps ecosystems healthy. Plants are **pollinated**, predators have prey, and water is clean. When animal species become endangered or extinct, however, these natural processes are disrupted. Plants die out, **invasive species** come in, and other animal species become endangered.

A trophic cascade happens when a top predator dies out or a new one is added. The Pacific sea otter, for example, has a great impact on coastal habitats. Scientists found that when Pacific sea otters were overhunted, the population of sea urchins, which were hunted by sea otters, became too large. This destroyed kelp forest habitats because too many sea urchins ate too much kelp. Later, when sea otters were protected, the kelp forests thrived.

*Wolves are a keystone species. These are species that have a huge effect on their ecosystem. Without them, the ecosystem would change dramatically or even collapse.*

*Elks overgrazed the land when gray wolves were hunted to near extinction. In 1974, protections for wolves were put in place in the U.S., and their ecosystems became healthy again.*

Since 1964, the International Union for the Conservation of Nature (IUCN) has maintained a Red List of Threatened Species. The Red List includes information on over 150,300 species, and researchers are working to include more.

Each year, scientists study and report the populations of animals and plants around the world. They classify species on the Red List by how vulnerable they are. Today, more than 42,100 species, or 28 percent of all assessed creatures, are threatened with extinction. The actual number could be much higher.

Critically endangered animals include the African forest elephant, hawksbill turtle, and the Bornean orangutan, but there are many more. Poaching, illegal wildlife trading, and **climate change** are all threats to animal survival. Earth is at risk of losing thousands of plant species as well, due to deforestation, disease, and other factors.

Many organizations, including World Wildlife Fund, are fighting to protect threatened species. Sanctuaries and rescue centers provide safe homes for endangered animals, while other organizations work to protect habitats.

African forest elephant

Bornean orangutan

Hawksbill turtle

After becoming extinct in the wild in 1972, the Arabian oryx was saved by a rescue operation. Today, there are about 1,200 living in the wild.

The Elephant Nature Park in Thailand is a sanctuary for endangered Asian elephants.

The western black rhino was declared extinct in 2011.

# What animal and plant species are already extinct?

Scientists estimate that more than 99 percent of species that ever lived on Earth are now extinct. Many extinctions were due to five naturally occurring mass extinctions. The first and second, 447 and 372 million years ago, killed marine organisms. The third and fourth, 252 and 208 million years ago, wiped out more marine species, insects, and land animals. Finally, about 66 million years ago, most dinosaurs and many other creatures were wiped out. Since then, many species have also become extinct due to human activity.

*Many animals went extinct after the last ice age, which ended around 10,000 years ago. Some scientists think that the woolly mammoth, for example, could not survive in Earth's warmer climate.*

*Scientists think the extinction event that wiped out most dinosaurs was caused by a giant* ***asteroid*** *hitting Earth.*

*In the 1800s, humans hunted the passenger pigeon to extinction.*

# Why do rainforests have so many different plants and animals?

Tropical rainforests are home to more than half of the world's land plants and animals. The Amazon rainforest alone has at least 40,000 plant species, more than 1,300 birds, and possibly 2.5 million types of insects. Scientists discover new species every year.

The climate and structure of tropical rainforests make this biodiversity possible. They are warm and wet all year, and the trees offer habitats and shelter for many animals and other plants. The four layers of a rainforest—emergent, canopy, understory, and floor—provide food and shelter for all different kinds of species.

*In the emergent layer, many birds, bats, and insects can be found.*

*The canopy has a rich mix of plants and supports animals such as monkeys, parrots, and reptiles.*

*In the understory, insects, snakes, and some mammals can be found.*

*The forest floor is full of insects and larger land animals, such as jaguars.*

## DID YOU KNOW?

**Rainforests are some of the world's oldest ecosystems. The Daintree Rainforest in northeastern Australia is believed to be 130 million to 180 million years old, making it the oldest rainforest in the world—tens of millions of years older than the Amazon rainforest.**

*The Daintree Rainforest is home to the cassowary (right), as well as tens of thousands of other species of birds, amphibians, reptiles, marsupials, insects, and plants.*

# How do fossils form?

Fossils are the remains of past organisms that are preserved in rock, soil, or **amber**. Scientists believe the oldest fossils ever found are forms of bacteria that lived underwater between 3.77 billion and 4.28 billion years ago.

Most fossils form as many layers of mud, sand, or other sediment cover and bury a dead organism. Gradually, these layers harden and become rock. Fossils can also be preserved in amber. This happens when tree resin, a sticky substance, traps and covers a living thing.

*Compared to other kinds of fossils, organisms preserved in amber are often more intact millions of years later.*

*Most fossils are found in sedimentary rock. As climate change causes glaciers to melt away from these rocks, scientists find more fossils.*

*Most plants cannot grow above the snow line, or the height above which snow covers the ground for most or all of the year. The tree line, lower than the snow line, is the height above which trees can no longer grow.*

# Can plants grow on top of mountains?

The tops of mountains are usually too rocky, high, and cold to support plant life. At high altitudes, air is colder, air pressure is decreased, and there is much less carbon dioxide. Plants need carbon dioxide to thrive. They also need to be pollinated by insects, but many insects cannot survive at high altitudes. This means that the highest mountains usually have no plant life at all.

Scientists have noticed, however, that trees and plants are growing higher up than they did before. This could be due to climate change. Warmer air, more rain, and an increase in carbon dioxide could all create better conditions for plants to grow.

# Why is Earth getting hotter?

Earth has been warming quickly since the middle of the 1800s, with the past 10 years being the warmest on record. This increase in Earth's average temperature, known as global warming, is part of climate change. Earth has had ice ages and warmer periods in its history, but these were caused by natural events such as volcanic eruptions. This period of climate change is because of human activity.

Greenhouse gases, such as carbon dioxide and methane, exist naturally in the atmosphere. They trap heat and help keep Earth warm and livable. In the 1800s, people began burning coal for factories and inventions such as steam trains. Forests were cleared for farming and to build cities and towns. This period was the **Industrial Revolution**.

Since then, the world has become more industrialized. Cars, planes, and many other machines and products using **fossil fuels** were invented. All this activity meant a massive increase in the carbon dioxide in the atmosphere.

## DID YOU KNOW?

**To learn how the atmosphere has changed over time, scientists can study the gases in air bubbles trapped in ice. They take samples in ice-covered places, such as Antarctica, drilling deep down into the ice. The deepest samples tested, in fact, contain information about the air 800,000 years ago! The amount of carbon dioxide in the samples helps scientists to determine how hot Earth was at times throughout history.**

*Ice core samples show that the amount of carbon dioxide in the atmosphere has increased drastically since the Industrial Revolution.*

*Scientists have found that carbon dioxide is increasing about 250 times faster due to human activities, compared to the rate of increase from natural sources.*

Scientists warn that Earth cannot get hotter than 2.7 degrees Fahrenheit (1.5 degrees Celsius) above what it was before industrial times. But experts say that in the next few years, the planet will surpass this limit. A few degrees may not seem like a lot, but global warming is already causing great harm.

Global warming is causing deadly heat waves and wildfires. These disasters kill people and wildlife, and destroy homes and habitats. As Earth heats up, glaciers and ice sheets are melting, and ocean levels are rising. Extreme weather events, such as floods and hurricanes, are becoming more frequent. Global warming is also causing more droughts, or long periods with little or no rain. This can lead to **desertification**.

People can take action to fight global warming. One of the most important things to do is to stop burning fossil fuels. Instead, people and governments can invest in renewable energy sources, such as the wind and the Sun. Protecting forests is also important because they absorb a lot of extra carbon dioxide from the atmosphere. People can speak up to their government representatives and push for change.

*Floods, droughts, and other extreme weather can make it difficult to grow crops. This means global warming is causing* ***food insecurity*** *in many places.*

A lot of carbon emissions come from manufacturing and selling goods such as clothing and electronics. Shopping second-hand is one way to reduce these emissions.

Wind turbines in Denmark provide about half the electricity for the whole country. They can be as tall as 919 feet (280 m) with 379 foot (116 m) long blades.

# Will Earth run out of water?

Earth has a huge amount of water. That means this valuable resource is not at risk for running out completely. However, not all people have equal access to clean fresh water.

The water cycle constantly recycles water as it moves from Earth to the atmosphere and back again. Global warming, however, affects the water cycle. Water is drying up in some areas, while ice is melting and not storing as much water as before. Flooding due to extreme weather pollutes water resources. Fighting global warming and protecting water from pollution are key for a healthy future.

*People cannot drink salty ocean water. Removing the salt through a process called desalination (above) can help bring fresh water where it is needed.*

*Less than 1 percent of Earth's water is accessible for human use. About 2 billion people do not have access to safe drinking water.*

*Restoring* ***mangrove*** *forests is important for protecting coastlines and communities from storms. Mangroves also absorb a lot of carbon dioxide and are home to a wide variety of wildlife.*

# Can damaged ecosystems be made healthy again?

Human activity has caused great damage to Earth's ecosystems. At the same time, many ecosystems such as forests and oceans are vital to help fight global warming because they take in carbon dioxide. Restoring ecosystems is possible and is important for all life on Earth.

In 2021, the United Nations (UN) established the Decade on Ecosystem Restoration. This global effort is focused on stopping the damage to ecosystems and restoring them. Forest ecosystems, for example, can be helped by replanting trees and allowing the forest to regrow naturally.

## DID YOU KNOW?

**About 75 percent of Costa Rica was tropical rainforest in the 1940s. Trees were cleared, however, for crops, timber, and cattle grazing. By 1987, its forest only covered about 25 percent of the country. Through reforestation and other efforts, rainforest now covers about 60 percent of Costa Rica.**

# What is coral bleaching?

Corals are tiny marine animals that live together in colonies. Many colonies form a coral reef. Naturally see-through, corals get their colors from a **symbiotic** relationship with different kinds of algae that live in their tissues. When ocean temperatures are too warm, corals expel the algae and turn white, or become bleached.

Global warming is the main cause of coral bleaching. Since corals use algae for food, without algae they are left vulnerable. Bleached coral can recover, but if the temperature stays too high and the algae does not come back, the coral will die. Once coral dies, reefs rarely come back.

*A temperature increase of just 2 to 4 degrees Fahrenheit (1 to 2 degrees Celsius) can cause the coral to stress and release the algae.*

*Scientists think that protecting whales and increasing whale populations can help tackle global warming.*

# Do animals affect global warming?

Cows and other livestock produce 14.5 percent of greenhouse gases in the world. Most is methane, which also comes from coal mining, natural gas systems, and landfills. There is less methane in the atmosphere than carbon dioxide. However, methane traps heat 25 times more powerfully than carbon dioxide.

Other animals, however, help fight global warming. Many spread seeds that grow into plants that absorb carbon dioxide. Others prey on plant-eating animals, keeping plant populations healthy.

## DID YOU KNOW?

**A whale captures about 33 tons (30 metric tons) of carbon dioxide over its lifetime. It also leaves waste that is food for tiny marine organisms called phytoplankton. Amazingly, phytoplankton consume about 40 percent of the world's carbon dioxide!**

*Cattle release most of the methane in their burps. Studies show that adding seaweed to a cow's diet helps reduce methane by 82 percent.*

# Glossary

**amber** (AM-ber): Fossilized tree resin, a thick, sticky substance produced in trees for protection

**asteroid** (AS-tuh-roid): Rocky space object that orbits the Sun

**atmosphere** (AT-muhs-feer): The layers of gases surrounding a planet

**climate change** (KLAHY-mit cheynj): The long-term change in Earth's usual climate

**composition** (kom-puh-ZISH-uhn): The way something is made up or put together

**denser** (denser): Having parts that are more closely packed together

**desertification** (dih-zur-tuh-fi-KEY-shuhn): A process by which land changes into desert

**epicenter** (EP-uh-sen-ter): Center point; starting point of an earthquake

**equator** (ih-KWEY-ter): An imaginary line around the middle of Earth

**fault** (fawlt): A long break in a rock formation

**food insecurity** (food in-si-KYOOR-i-tee): Not having access to enough nutritious food

**fossil fuels** (FOS-uhl FYOO-uhls): Fuels, such as coal, made from ancient plant or animal remains

**friction** (FRIK-shuhn): The force that happens when two objects rub against each other

**galaxy** (GAL-uhk-see): A large group of stars and their solar systems, dust, and gas

**geocorona** (jee-oh-kuh-ROH-nuh): The outermost layer of Earth's atmosphere, made mostly of hydrogen

**Industrial Revolution** (in-DUHS-tree-uhl rev-uh-LOO-shuhn): The period in which humans moved from an agricultural-based society to one based on industry, or the manufacturing of goods

**invasive species** (in-VEY-siv SPEE-seez): A species not native to an ecosystem that causes harm

**magnetic field** (mag-NET-ik feeld): An area around a magnetic body in which the force of magnetism is at work

**mangrove** (MAN-grohv): A tropical tree that grows in tidal areas

**meteoroids** (MEE-tee-uh-roids): Rocky space objects smaller than asteroids

**organisms** (AWR-guh-niz-uhms): Living things

**pollinated** (POL-uh-neyted): Transferred pollen from one plant to another to produce new plants

**Precambrian** (pree-KAM-bree-uhn): Relating to the earliest time in Earth's history

**radiation** (rey-dee-EY-shuhn): Energy that travels from a source in waves or particles

**radioactive elements** (rey-dee-oh-AK-tiv EL-uh-muhnts): Substances that give off radiation

**sedimentary** (sed-uh-MEN-tuh-ree): Rocks created from the buildup of sediment over time

**shock waves** (shok weyvs): Movements of high air pressure or temperature, caused by events such as explosions and earthquakes

**solar wind** (SOH-ler wind): The fast-moving stream of particles flowing outward from the Sun

**stalagmite** (stuh-LAG-mahyt): Rock formation that rises as a column from a cave floor, formed from material that drips from the ceiling

**symbiotic** (sim-bahy-OT-ik): A close relationship between organisms that benefits both

**tabular** (TAB-yuh-ler): Wide and flat, like a tabletop

**tectonic plates** (tek-TON-ik pleyts): Large pieces of Earth's crust that float on the mantle

**trench** (trench) A long, deep cut in Earth's surface

# Index



# Comprehension Questions

1. **About ________ of all earthquakes happen in the Ring of Fire.**
   a) 75 percent
   b) 90 percent
   c) 95 percent

2. **True or False:** About 5 million people do not have access to safe drinking water.

3. **________ blow from east to west near the equator.**
   a) Trade winds
   b) Westerlies
   c) Kabatic winds

4. **True or False:** Lake Baikal is Earth's deepest lake.

5. **True or False:** Scientists have found Earth's oldest rocks in Québec, Canada.

6. **The International Space Station orbits Earth in the ________.**
   a) thermosphere
   b) mesosphere
   c) stratosphere

7. **The ________ is the saltiest ocean.**
   a) Pacific
   b) Indian
   c) Atlantic

8. **True or False:** Carbon dioxide is increasing about 100 times faster due to human activity, compared to natural increases.

1. b, 2. False, 3. a, 4. True, 5. True, 6. a, 7. c, 8. False

# About the Author

Cynthia O'Brien's curiosity about the world led her to be a writer for young readers. She has written over 80 books covering a wide range of subjects.

## Crabtree Publishing

**crabtreebooks.com** **800-387-7650**





In Canada: We acknowledge the financial support of the Government of Canada through the Canada Book Fund for our publishing activities.

| | |
|---|---|
| Hardcover | 978-1-0398-8090-0 |
| Paperback | 978-1-0398-8451-9 |
| Ebook (pdf) | 978-1-0398-8210-2 |
| Epub | 978-1-0398-8330-7 |

Printed in the U.S.A./CP2025

**Published in Canada**
**Crabtree Publishing**
616 Welland Avenue
St. Catharines, Ontario
L2M 5V6

**Published in the United States**
**Crabtree Publishing**
347 Fifth Avenue
Suite 1402-145
New York, NY 10016

**Written by:** Cynthia O'Brien
**Series development:** Janine Deschenes
**Editorial coordinator:** Kathy Middleton
**Editor:** Janine Deschenes
**Proofreader:** Melissa Boyce
**Art direction and layout:** Katherine Kantor and Margaret Salter
**Photo research:** Katherine Kantor
**Production manager:** Candice Campbell

**Library and Archives Canada Cataloguing in Publication**
Available at the Library and Archives Canada

**Library of Congress Cataloging-in-Publication Data**
Available at the Library of Congress

**Images:**
NASA: p. 25 (bottom right)
JPL: p. 4–5 (Earth), 9 (binocular inset)
NOAA: p. 23
Shutterstock
Knumina Studios: p. 25 (top)
U.S. Fish and Wildlife Service: p. 33 (bottom)
Wikimedia Commons
Dave Pape: p. 19
Helle Astrid Kjær: p. 38 (bottom left)
Laurent Augustin: p. 38 (center)
US Geological Survey, outline of Williston Basin added by Geologyguy: p: 6
All other images by Shutterstock